Intermittent Fasting Diet

The Intermittent Fasting Cookbook - Delicious Recipes for the Intermittent Diet

Lindsay Parsons

Copyright © 2013 Lindsay Parsons
All rights reserved.

Table of Contents

INTRODUCTION 1

INTERMITTENT FASTING DIET RECIPES 10

Intermittent Fasting Diet Breakfast Recipes 10
- Breakfast Casserole 10
- Healthy Breakfast Burrito 12
- Mexican Style Eggs "Huevos Rancheros" 14
- Mexican Breakfast Casserole 16
- Savory Hash Browns 18
- Squash, Zucchini and Eggs 20
- Tomato Spinach Eggs 22
- Whole Grain Hot Cereal with Cherries 24
- Whole Wheat Pancakes with Apples 26
- Zucchini Frittata 28

Intermittent Fasting Diet Dinner Recipes 30
- Balsamic Turkey Meatloaf 30
- Buffalo Chicken with Slaw 32
- Edamame and Grilled Salmon 34
- Grilled Chicken Tostadas 35
- Italian Chicken 37
- Oriental Turkey Burgers 40
- Shepherd's Pie 43
- Shrimp Scampi 46
- Vegetable Pot Pie 48

Intermittent Fasting Diet Light Snack Recipes 50
- Apple and Turkey Ham Salad 50
- Baked Potatoes Twice 51
- Broccoli Cheese Soup 53
- Cauliflower Soup 56
- Greens with Baked Beans 58
- Maple Flavored Sweet Potato Fries 60
- Nutty Cucumber Mango Rice Salad 62
- Open Face Tomato and Mozzarella Herb Sandwich 63

Orange Stir Fry Vegetables 65
Parsley Mint Roasted Carrots 67
Quinoa with Herbs.. 69
Spicy Tomatoes and Green Beans 71
Spinach Salad with Pomegranate Dressing.. 72

Introduction

What is the intermittent fasting diet?

This is a diet in which you eat during specified time frames. There are two popular versions of this diet. One being a day to day eat and fast where you eat on day one, fast on day two, and repeat until the desired weight is lost. The other version is a daily fasting, where you eat for a six to eight hour and then fast the rest of the day.

Starting out on the intermittent fasting diet

Please be aware that results from the intermittent diet vary from person to person. Much of the variance depends upon the build of the body, how much fat, and weight need to be lost and how they eat during the diet. Other factors that influence weight loss are lifestyle (do you smoke? drink? eat excessive junk food?), insulin resistance, exercise or not, and work. All these things work towards either making it difficult to lose the weight or to lose it fast. No two people are alike, even if they each desire to lose the same amount of weight. Keeping this in mind helps you to tailor the diet to your own needs.

Intermittent means eating and fasting in chunks of time. You may need to adjust this as you go along, to help deal with other health issues, to speed things along or to help improve health. Be ready to make adjustments and learn how to bend with the changes. Keeping a positive attitude will carry you a long ways in having good success in this diet.

This is a great diet to start for good weight loss. When the weight and fat are gone you can go on a maintenance, where you may still continue to fast, but more on a less restriction.

Other Issues Helped By Intermittent Fasting

If you suffer from any type of insulin or blood sugar problems, the metabolism is improved by fasting, eating in this manner. Eating through a fasting manner helps the body to lower inflammation rates, helps to improve blood pressure, helps to release stress, and helps to boost the immune system. If the immune system is boosted, the body is able to fight off other illnesses and keep you healthy and strong. Because this diet helps also to increase metabolism (especially if you eat the right food) you will have more energy to exercise and your body will have more energy to digest and disperse the foods you eat.

One of the biggest reasons people gain so much weight is they feel the need to eat all the time. This constant eating causes people to grab for the fast, convenient, and high in sugar and salt foods. These foods are responsible for putting a massive amount of weight on those who gorge. The intermittent fasting diet stops this binge eating and helps to creating a habit of eating during the best times of the day.

The Basics of the Intermittent Fasting Diet

The intermittent fasting diet is an extremely flexible diet plan. There are no set rules for doing it other than having a nice solid block of time to fast. As mentioned above one of the methods of fasting involves taking a day or longer of no food, and doing this a couple of times a week. For obvious reasons, this mode is a lot tougher to deal with, as going a solid day without food may be impossible for some people, especially for those with blood sugar issues.

Daily intermittent fasting is better because it does allow for the intake of food on a daily basis. But it also means for a strict window of daily fasting too and this is what helps to facilitate the weight loss. Basically you eat during a six to eight hour time and fast the remainder.

Your lifestyle will directly affect the effectiveness of this diet. Because the diet is flexible, there are no set foods or meals to eat during the six to eight hour window, just to eat. If you are in the habit of consuming a lot of carbohydrates, (sugars, white flours, and basically food with no nutritious value) then your weight loss may not happen or may be very slow. If your lifestyle is very sedentary, it will take longer to lose the weight.

Here is the thing, if you clean up your eating habits and eat foods packed with nutrition, it will give your body the energy to burn to move about more. You will want to exercise and move about more. Your digestive system will also work more efficiently, digesting in a time and manner that will get rid of the fat and calories.

Choosing the Daily Intermittent Fasting

This book is geared to offer recipes for the daily plan. It is easier to follow the daily fasting routine and can be developed into a habit, which will help it to be easier to do. You will break bad habits of needing to binge or gorge on food, because you know that after your six to eight hours, you simply will not eat. Your body will be able to adjust to this much easier, because you will have food in your body daily. If you eat the right foods, your body will have nutrients in which to work to help keep you healthier.

Snacking is the downfall of many who may eat well during the meals but find themselves reaching for foods void of nutrients. Unfortunately, these types of foods are highly addictive, the more we consume them the more our bodies want it. But it is also a habit that is fairly easy to break if you have the will power to do so. This diet stops the constant snacking.

By fasting on a daily basis, you will be more aware of your body. The hope is you will be aware of the foods you eat during the feeding hours, and will choose to consume healthier foods and snacks.

Be Aware Of Issues

If you do not consume enough calories during the feeding window, you can run the risk of reaching a weight loss plateau. This occurs when the body is too restricted from food (remember what we discussed about "starvation mode" above?) This can be avoided by eating the right foods. If you consume junk foods, then your body will not have substantial energy to keep going during the fasting period. If you eat a balance of good complex carbohydrates, proteins, and nutrients, your body can easily sustain the diet and the body during the fasting.

Some people will not be able to do the intermittent fasting diet. Some people have a greater need for more calories and they simply will not perform well unless they are consuming these calories. It is wise to have a physical and make sure your body can handle such a diet. Go over your choices with your health care provider and let them help you to decide whether or not this diet is right and healthy for you.

Making the Intermittent Diet a Success

This diet can be a great success if you do it right. It does take work and dedication though. First thing is to eat right. Choose foods that are healthy and whole during your feeding and avoid junk foods altogether. Eat a good breakfast each morning that will fuel your body to keep it going during the day. Choose healthy snacks. In the sample 5 day meal plan we suggest to eat fruit and nuts to snack. It is okay to drink fruit juice and eat leftovers or a small meal if you would rather. The point is to consume foods like lean meats, fruits, vegetables, and whole grains that help to give the body all the nutrients it needs to function at optimum levels.

Once the feeding window closes, do not consume any more food until the next day. It is okay and encouraged, though, to drink plenty of water throughout the entire day. Water helps to facilitate weight loss and helps to cleanse the body of impurities and toxins. Try drinking water throughout the feeding window as well. It helps with digestion too.

Take up an exercise routine. If your body is moving around it helps to burn more calories. Exercise also helps the body to release endorphins, and these are nature's way of giving you a natural high. Exercise is addictive too, the more you do it the more you will want to do it. The toughest part is starting. Even if you only work out three times a week for thirty minutes each time, you are giving your body a greater chance of fat and weight loss by doing so.

Sample 5 Day Meal Plan

The three meals here are smaller portions than a regular meal. The meals are to be eaten in a 6 to 8 hour time frame, with a 16 to 18 time frame of fasting. During the fasting time, you can have water. It is okay to have more with the meals if you are hungry. Try having a salad with lunch and supper if needed. Drink plenty of water during the day too.

Day One

Breakfast - Tomato Spinach Eggs
Snack - Nuts
Lunch - Edamame and Grilled Salmon
Snack - Fruit
Supper - Apple and Turkey Ham Salad

Day Two

Breakfast - Whole Grain Hot Cereal with Cherries
Snack - Nuts
Lunch - Balsamic Turkey Meatloaf
Snack - Fruit
Supper - Broccoli Cheese Soup

Day Three

Breakfast - Savory Hash Browns
Snack - Nuts
Lunch Buffalo Chicken with Slaw
Snack - Fruit
Supper - Open Face Tomato and Mozzarella Herb Sandwich

Day Four

Breakfast - Mexican Breakfast Casserole
Snack - Nuts
Lunch - Shrimp Scampi
Snack - Fruit
Supper - Spinach Salad with Pomegranate Dressing

Day Five

Breakfast - Healthy Breakfast Burrito
Snack - Nuts
Lunch - Italian Chicken
Snack - Fruit
Supper - Baked Potatoes Twice

Intermittent Fasting Diet Recipes

Intermittent Fasting Diet Breakfast Recipes

Breakfast Casserole

This makes a perfect brunch because it is a hearty and filling casserole of eggs, cheese, tomatoes, and English muffins. This recipe needs to be prepared the night before. Makes 8 servings.

What You'll Need:

4 English muffins (halved, toasted)
4 scallions (cut into long bite-sized pieces)
4 eggs plus 3 egg whites
3 cups of milk (low fat)
2/3 cup of cheddar cheese (extra sharp, shredded, divided)
3/4 cup of deli ham (torn, thin slices)
1/2 cup of tomatoes (no oil, sundried, sliced)
1 tablespoon of Dijon mustard

Salt and pepper

How to Make It:

Spray a 2-quart baking dish with cooking spray. Layer the bottom of the pan with the 4 halved and toasted English muffins and the 3/4 cup of torn thin sliced deli ham, making sure to lay it out evenly. Next layer it with the 4 scallions cut into bite-sized pieces, 1/2 cup of sliced sundried tomatoes, and 1/3 cup of shredded extra sharp cheddar cheese. In a bowl, crack the 4 eggs and add the 3 egg whites and beat with a whisk, then combine with the 3 cups of low fat milk, tablespoon of Dijon mustard and dashes of salt and pepper. Pour the egg mixture over the layered casserole in the baking dish. Add the remaining 1/3 cup of extra sharp cheddar cheese on top. Cover tightly with plastic wrap and refrigerate overnight. Next morning, preheat the oven to 350 degrees Fahrenheit. Remove the plastic wrap and cook the casserole (with a baking sheet under it) for 60 minutes. Remove from oven and let it sit for at least 10 minutes to set and cool before serving and enjoying.

Healthy Breakfast Burrito

Mornings are the time to refuel for the day. Start the day right with a breakfast burrito that is as healthy as it is delicious. Makes 4 servings.

What You'll Need:

4 tortillas (whole wheat, burrito)
4 eggs
4 egg whites
1 avocado (cubed)
1 cup of onions (diced)
1 cup of black beans (cooked, rinsed)
3/4 cup of tomatoes (diced)
1/2 cup of bell peppers (red, seeded, diced)
1/3 cup of pepper Jack cheese (shredded)
1/4 cup of sour cream
1/4 cup of salsa
2 teaspoons of canola oil
1/4 teaspoon of red pepper flakes
Salt and pepper
Hot sauce

How to Make It:

Add the 2 teaspoons of canola oil to a skillet on medium

high heat. Sauté the 1/2 cup of seeded, diced red bell peppers and 1/2 cup of diced onions. Stir in the cup of cooked, rinsed black beans and the 1/4 teaspoon of red pepper flakes. Cook for a couple of minutes to warm, and then add dashes of salt and pepper. Put the contents into a bowl and set aside. In a separate bowl crack the 4 eggs and add the additional 4 egg whites and whisk. Add the 1/3 cup of shredded pepper Jack cheese. Spray the same skillet with cooking spray, heat to medium and scramble the eggs until done. In a separate non-stick skillet, heat to medium, and warm each tortilla on each side for about 30 seconds. Next, to build the burrito, add 1/4 of the sour cream and 1/4 of the salsa followed by 1/4 of the black beans and topped with 1/4 of the eggs and then season with extra salt, pepper, and hot sauce. Roll up and serve. Do this with each one.

Mexican Style Eggs "Huevos Rancheros"

If you love Mexican food, you will love your breakfast fixed fiesta style. Makes 4 servings.

What You'll Need:

4 eggs
4 tortillas (corn, 6 inch, warmed)
1 can of black beans (15.5 oz., drained, rinsed)
1 jalapeno pepper (minced)
1 1/2 cups of tomatoes (fine chopped)
1/2 cup of onions (fine chopped)
1/2 cup of feta cheese (crumbled)
1/2 cup of water (warm)
1/4 cup of cilantro (fresh chopped)
2 tablespoons of olive oil (plus 2 teaspoons, extra virgin)
1 teaspoon of garlic (minced)
1 teaspoon of cumin (ground)
1/2 teaspoon of hot sauce
Salt and pepper

How to Make It:

Mix the 1 jalapeno pepper (minced), 1 1/2 cups of tomatoes (fine chopped), 1/2 cup of onions (fine chopped), 1 teaspoon of garlic (minced), 1 teaspoon of cumin (ground), 1/2 teaspoon of hot sauce, and dashes of salt and pepper together to make salsa. Pour 2 teaspoons of olive oil into a skillet and heat to medium low. Pour in the "salsa" mixture and stir for a couple of minutes until it thickens. Pour the salsa in a bowl and set to the side. Add the can of drained, rinsed black beans along with 1/2 cup of warm water and another dash of salt into the skillet. Cover, turn heat to low and simmer while preparing the rest of the eggs. Add the 2 tablespoons of extra virgin olive oil to another skillet and heat to medium. Crack the eggs, one at a time to make 4 fried eggs, or sunny side up. Season with salt and pepper. Warm the 4 tortillas by placing on a plate with a damp paper towel on top and microwave for about 20 seconds. Next, place a tortilla on 4 plates. Equally divide the beans on top of the 4 tortillas. Add a fried egg to each one. Add a spoon of salsa on top of each egg, and then divide the 1/2 cup of feta cheese crumbles on top of the salsa. Garnish with the 1/4 cup of fresh chopped cilantro and the rest of the salsa. Serve immediately.

Mexican Breakfast Casserole

Here is a casserole filled with the spiciness of chili and cilantro, delicious and filling. Makes 6 servings.

What You'll Need:

4 cups of tortilla chips (baked, divided)
4 eggs plus 6 extra egg whites
1 can of green chilies (chopped, drained)
1/2 cup of cheddar cheese (sharp, shredded - divided)
1/2 cup of pepper Jack cheese (shredded - divided)
1/2 cup of salsa (green Verde)
1/4 cup of skim milk
1 tablespoon of cilantro (fresh chopped plus more for garnishment)
3/4 teaspoon of ancho chili powder
Dollops of sour cream
Salt and pepper

How to Make It:

Prep: Preheat the oven to 375 degrees Fahrenheit. Spray a 2 quart baking dish with cooking spray.

Crumble the 4 cups of baked tortilla chips (large crumbles) and lay 2 cups of chips in the bottom of the baking dish. In a bowl, crack the 4 eggs and add the 6 egg whites and beat with a whisk. Add the 1/4 cup of skim milk, 3/4 teaspoon of ancho chili powder, and dashes of salt and pepper and stir. Mix in the can of chopped drained green chilies, 1/4 cup of shredded sharp cheddar cheese, 1/4 cup of shredded pepper Jack cheese, and the tablespoon of fresh chopped cilantro. Pour the mixture over the baked tortilla chips in the baking dish. Place in hot oven and bake for about 22 minutes, until the eggs are set. Pull out of oven and sprinkle the remaining 1/4 cup of shredded sharp cheddar cheese and the 1/4 cup of shredded pepper Jack cheese and place back in the oven for 10 more minutes. Pull from oven, turn heat off, and allow sitting for another 10 minutes. Serve with a spoon of green Verde salsa, dollop of sour cream, and a garnishment of cilantro leaf.

Savory Hash Browns

All you need to do is cook up and egg and have a piece of whole grain toast and you are set for a meal. Makes 4 servings.

What You'll Need:

2 potatoes (Yukon gold, washed, grated - with skins)
2 scallions (chopped)
1 parsnip (peeled, grated)
2 tablespoons of parsley (minced flat leaf)
1 tablespoon of olive oil (extra virgin, divided)
Salt and pepper

How to Make It:

Toss the 2 grated potatoes with the grated parsnip, add the 2 chopped scallions, greens and all. Season with dashes of salt and pepper. Pour the 1/2 tablespoon of extra virgin olive oil into a skillet and heat to medium. Stir in the grated potatoes, parsnips, and scallions, tossing to coat with oil, then press down into the skillet, once in a while, run the spatula under the mixture to prevent sticking. Cook until crispy brown for around 10 minutes. Flip the mixture out onto a large dinner plate. Add the remaining 1/2 tablespoon of olive oil, return skillet to heat, then replace the mixture, uncooked side down to crisp the other side, another 10 minutes. Serve hot.

Squash, Zucchini and Eggs

This is a great summer meal using fresh squash if possible. Makes 6 servings.

What You'll Need:

6 eggs
4 scallions (sliced thin, greens separated out)
3 squash (grated)
3 zucchini (grated)
1 jalapeno (seeded, minced)
1/4 cup of cheddar cheese (sharp white, shredded)
1/4 cup of pepper jack (shredded)
3 tablespoons of parsley (fresh chopped)
2 tablespoons of olive oil (extra virgin)
1 tablespoon of butter
1 tablespoon of salt
1/4 teaspoon of nutmeg (ground)
Salt and pepper

How to Make It:

Toss the 3 shredded squash and the 3 grated zucchinis with a tablespoon of salt while they rest in a colander for 35 minutes. Using a paper towel, squeeze the squash and zucchini.

Preheat the oven to 375 degrees Fahrenheit. Place an oven proof (cast iron works well) skillet on the stove on medium high heat. Pour in the 2 tablespoons of extra virgin olive oil. Reserve 3 tablespoons of the greens from the 4 scallions and put the remainder of the greens and all the whites into the heated oil along with the seeded minced jalapeno and sauté. Toss in the grated squash and zucchini, stir, and toss for about 7 minutes. Add the 3 tablespoons of fresh chopped parsley, 1/4 teaspoon of ground nutmeg, and dashes of salt, pepper, and stir, cooking for another minute. Remove skillet from the stove and sit for 5 minutes away from the heat. Next, pat the squash, zucchini mixture down, then with the back of a serving spoon make 6 indentions, spaced evenly over the squash and zucchini. Place 1/2 of a teaspoon of butter into each of the 6 indentions. Carefully, crack an egg in a cup, then pour right into and indention, with all 6 eggs. Sprinkle dashes of salt and pepper over the eggs. Add the remaining 1/4 cups of shredded sharp cheddar cheese and pepper Jack cheese, evenly over the top. Carefully place the skillet in the hot oven and bake for about 11 minutes. Garnish with the 3 tablespoons of the chopped green scallions and serve immediately.

Tomato Spinach Eggs

This is a delicious way to get protein and vegetables first thing in the morning, with this savory eggs Benedict recipe. Makes 4 servings.

What You'll Need:

8 cups of spinach (fresh baby)
4 slices of tomato (large slices)
4 eggs
2 English Muffins (split in half)
1/2 cup of onions (thin sliced)
1/3 cup of Canadian bacon (chunked)
1/4 cup of vinegar (white distilled)
2 tablespoons of mayonnaise
1 tablespoon of water (warm)
1 tablespoon of olive oil
2 teaspoons of mustard
1 teaspoon of lemon juice
Dash of cayenne pepper
Pepper

How to Make It:

Make the sauce by combining the 2 tablespoons of mayonnaise, 1 tablespoon of water (warm), 2 teaspoons of mustard, 1 teaspoon of lemon juice, and dash of cayenne pepper with a whisk.

Next, start the Benedict eggs by adding several inches of water to a large saucepan. Pour in the 1/4 cup of white distilled vinegar and turn the heat to medium.

Set a nonstick frying pan on medium high, add the tablespoon of olive oil, 1/3 cup of Canadian bacon chunks, and the 1/2 cup of onions, and cook until heated through. Stir in the 8 cups of fresh baby spinach, take the frying pan off the heat, and continue stirring for a couple of minutes until the leaves wilt. Sprinkle pepper and toss.

Pop the English muffins into a toaster to toast lightly on all sides. Set them on a serving platter and top with a slice of tomato. Add 1/4 of the Canadian bacon mixture on top of each tomato slice.

Next, cook one egg at a time, buy cracking into a small dish, then pouring into the simmering vinegar water. Cook for about 4 minutes. Remove and place on top of the Canadian bacon on the English muffin halves. Do this with all 4 eggs. Spoon the hollandaise sauce over the top and serve hot.

Whole Grain Hot Cereal with Cherries

There is nothing heartier than a bowl of hot whole grain cereal first thing in the morning. You will enjoy this meal with the aroma and flavor of fruit making it a delightful meal. Makes 4 servings.

What You'll Need:

5 cups of water
1/2 cup of rice (wild)
1/2 cup of oats (steel-cut)
1/2 cup of wheat cereal (cream of wheat)
1/4 cup of pearl barley
1/4 cup of cherries (dried)
1 cinnamon stick
1 1/2 tablespoons of brown sugar (packed)
1/2 teaspoon of orange zest
1/4 teaspoon of salt
Walnuts (chopped)
Butter
Milk

How to Make It:

The evening before add the 5 cups of water, 1/2 cup of rice (wild), 1/2 cup of oats (steel-cut), 1/2 cup of wheat cereal (cream of wheat), 1/4 cup of pearl barley, 1/4 cup of cherries (dried), 1 cinnamon stick, 1 1/2 tablespoons of brown sugar (packed), 1/2 teaspoon of orange zest, and 1/4 teaspoon of salt and stir in a large sauce pan. Place the cover and let sit on the stove with the heat off over night. The morning of breakfast, turn the stove on high and bring to a boil, then turn it down on low to simmer, cover on for 20 minutes. Keep the cover on, turn the stove off and let it sit for another 5 minutes. Serve in bowls and garnish with chopped walnuts, butter and milk if desired.

Whole Wheat Pancakes with Apples

Pancakes are always fun to cook and eat. You cannot go wrong with this recipe, which uses whole-wheat flour to give you the benefit of whole grains, and the goodness of fresh apples. Makes 6 servings.

What You'll Need:

1 cup of buttermilk (low fat)
3/4 cup of skim milk
3/4 cup of apples (cored, diced)
3/4 cup of flour (all-purpose)
3/4 cup of flour (whole-wheat)
2 eggs
6 tablespoons of maple syrup
1 tablespoon of honey
2 teaspoons of baking powder
1/2 teaspoon of baking soda
1/4 teaspoon of salt

How to Make It:

Prep: Preheat the oven to 250 degrees Fahrenheit.

In a bowl, combine the 3/4 cup of flour (all-purpose), 3/4 cup of flour (whole-wheat), 2 teaspoons of baking powder, 1/2 teaspoon of baking soda, and 1/4 teaspoon of salt. Crack the 2 eggs and beat with a whisk in a cup. In a separate bowl, combine the 1-cup of buttermilk (low fat), 3/4 cup of skim milk, beaten eggs, and the tablespoon of honey. Gradually add the dry ingredients, do not over stir.

Next, place the diced apples in a microwave safe dish, cover with plastic wrap, and microwave on normal for 2 minutes to soften.

Turn the heat to medium on a non-stick skillet or griddle. Ladle out about a fourth a cup of batter onto the hot surface. Spoon a couple of apples over the top, flip after a couple of minutes. Repeat until all the batter and apples are gone. Drizzle with the maple syrup or your favorite syrup over the top, or sprinkle cinnamon and sugar over the top.

Zucchini Frittata

This delicious breakfast would make a good dinner choice too, as it's healthy and filling, full of feta cheese, zucchini, potatoes and turkey bacon. Makes 4 servings.

What You'll Need:

4 eggs + 2 egg whites
2 strips of turkey bacon (cooked, crumbled)
1 zucchini (grated and dried with a towel)
1 cup of potatoes (russet, cubed)
1/2 cup of feta cheese
1/2 cup of onion (chopped fine)
2 tablespoons of cilantro (fresh chopped)
1 tablespoon of olive oil
3/4 teaspoon of salt
1/2 teaspoon of garlic (minced)
1/4 teaspoon of hot sauce

How to Make It:

Add the 1 cup of cubed potatoes to a saucepan and cover with water, bring to a boil on high heat, then turn down to medium high. Cook for about 7 minutes or until the potatoes are tender enough to eat. Remove from heat, drain the water and place in a bowl. Using a paper towel, dry the potato cubes.

In a bowl, add the 4 eggs and 2 egg whites and beat with a whisk. Stir in the cup of cilantro, 3/4 teaspoon of salt, and 1/4 teaspoon of hot sauce.

Turn the on the oven broiler to high.

Place an ovenproof skillet on the stove (about a 10 inch size) and turn to medium high heat. Add the tablespoon of olive oil and sauté the 1/2 cup of fine chopped onion and the 1/2 teaspoon of minced garlic. Stir in the grated zucchini and cook for another 5 minutes. Stir in the cooked potato cubes, browning them for about 4 minutes. Next, pour the whisked egg mixture over the potatoes and zucchini. Place the skillet back on medium heat, lifting the edges to allow the egg to run, for a couple of minutes. Next, sprinkle the 2 strips of crumbled turkey bacon and the 1/2 cup of feta cheese over the top and place under the broiler for 5 minutes. Serve hot.

Intermittent Fasting Diet Dinner Recipes

Balsamic Turkey Meatloaf

If you are a meatloaf lover you will enjoy this different twist for meatloaf, which is a bit healthier than the beef counterpart. Makes 8 servings.

What You'll Need:

1.5 pounds of ground turkey
1 zucchini (fine diced)
1 bell pepper (red fine diced)
1 bell pepper (yellow fine diced)
1 egg
1 cup of bread crumbs
3/4 cup of ketchup (divided)
1/4 cup + 2 tablespoons of balsamic vinegar
1/4 cup of Parmesan cheese (grated)
1/4 cup of Romano cheese (grated)
1/4 cup of parsley (fresh chopped)
2 tablespoons of olive oil (extra-virgin)
1 tablespoon of thyme (fresh fine chopped)
2 1/2 teaspoons of garlic (minced)
1/2 teaspoon of red pepper flakes
Salt and pepper

How to Make It:

Prep: Preheat oven to 425 degrees Fahrenheit. Line a 9x5 inch loaf pan with foil.

Add the 2 tablespoons of extra virgin olive oil to a skillet on high heat and sauté the fine diced zucchini, red and yellow bell peppers, 2 1/2 teaspoons of minced garlic and dashes of salt and pepper for about 5 minutes. Set aside.

Crack the egg in a bowl and beat with a whisk, and stir in the 1/4 cup of fresh chopped parsley and the tablespoon of fresh fine chopped thyme. Add the 1.5 pounds of ground turkey, breaking it up with your hands, along with the cup of breadcrumbs, 1/4 cup of grated Parmesan cheese, 1/4 cup of grated Romano cheese, 1/2 cup of ketchup, 2 tablespoons of balsamic vinegar, and the zucchini and bell peppers. Mix with bare hands and mold into a loaf. Add to the lined loaf pan. Make the sauce for the topping by mixing the 1/4 cup of ketchup with the 1/4 cup of balsamic vinegar and the 1/2 teaspoon of red pepper flakes, and dashes of salt and pepper. Stir with a whisk, and then pour over the top of the meat loaf. Cook for 1 hour and 15 minutes, until the internal temperature of the meatloaf reaches 165 degrees Fahrenheit with a meat thermometer.

Buffalo Chicken with Slaw

Buffalo chicken is always associated as an appetizer but here it is a delicious main meal with a side of fresh homemade slaw. Makes 4 servings.

What You'll Need:

4 chicken breast halves (boneless, skinless, cut into strips)
4 cups of cabbage (shredded)
2 cups of buttermilk
2 cups of carrots (grated)
2 cups of bread crumbs (fine)
1 cup of celery (thin sliced)
1/2 cup of canola oil
1/2 cup of mayonnaise
1/2 cup of sour cream
1/2 cup of blue cheese (crumbles)
2 tablespoons of hot sauce (divided)
Salt and pepper

How to Make It:

Combine the 2 cups of buttermilk with 1 tablespoon of hot sauce and dashes of salt and pepper. Put the 4 boneless, skinless chicken breast halves cut into strips into a shallow dish. Pour the buttermilk mixture over the chicken, cover and refrigerate for 60 minutes. Combine the 1/2 cup of mayonnaise with the 1/2 cup of sour cream and the 1/2 cup of blue cheese crumbles in a blender or food processor until nice and lump free (this is the dressing). Using a whisk, add the remaining tablespoon of hot sauce and mix. In a bowl, add the 4 cups of shredded cabbage, 2 cups of grated carrots, with the 1 cup of thin sliced celery and toss. Pour 3/4 cup of the dressing over the cabbage mixture and toss to coat all. There will be 1/4 cup of dressing left over for a dipping sauce.

Add the 2 cups of fine bread crumbs to a shallow dish. Shake each chicken strip from the marinade and roll in the bread crumbs. Pour the 1/2 cup of canola oil into a skillet and turn to medium high heat. Fry each coated chicken strip for 4 minutes, turn and cook another 4 minutes, until all the chicken is cooked.

Serve with a side of slaw and dip in the dressing.

Edamame and Grilled Salmon

It is hard to beat salmon in terms of nutrition and flavor. This delicious meal is savory to the palate the filling. Makes 4 servings.

What You'll Need:

4 salmon fillets (skin on)
2 scallions (fine chopped)
1 1/3 cup of edamame (cooked)
1/4 cup of cilantro leaves (fresh fine chopped)
2 teaspoons of canola oil
2 teaspoons of lime juice
2 teaspoons of soy sauce
2 teaspoons of honey
1 teaspoon of ginger (grated)
1/4 teaspoon of sesame seeds (black)
Salt and pepper
lime wedges (for garnish)

How to Make It:

Prep: Preheat the grill to medium high. Rub canola oil on the grates.

Mix the 2 fine chopped scallions with the 1/4 cup of

fresh fine chopped cilantro leaves, 2 teaspoons of canola oil, and the teaspoon of grated ginger. Dash salt and pepper and toss. Cut into the middle of the skins of the salmon fillets, making 2 slits about three inches in length from top to bottom, cutting halfway into the salmon. Do so with each fillet, and evenly spoon the scallions and cilantro into each slit. Salt and pepper the rest of the salmon fillets. In a cup, mix the 2 teaspoons of lime juice, 2 teaspoons of soy sauce, with the 2 teaspoons of honey with a whisk. Gently set each salmon fillet on the grill with the skin / herbs side facing up. Grill for about 3 1/2 minutes. Flip the salmon, brush the top with the lime juice sauce mixture and grill for another 3 1/2 minutes. Place cooked salmon fillets on a serving platter; evenly sprinkle the 1/4 teaspoon of black sesame seeds over the tops. Garnish with the lime wedges and serve with the 1 1/3 cup of cooked edamame in a serving dish.

Grilled Chicken Tostadas

This is a healthy meal made with tasty seasoned chicken breasts and a variety of other savory flavors. Makes 4 servings.

What You'll Need:

4 tortillas (flour, 8 inch)
2 chicken breasts (boneless, skinless, cut into bite-sized pieces)
1 pound of tomatillos (husked and rinsed)
4 lime wedges
1 chipotle chili in adobo sauce (chopped coarse)
2 cups of romaine lettuce (shredded)
1/3 cup of feta cheese (crumbled)
1/4 cup of lime juice
4 tablespoons of onions (fine chopped)
2 tablespoons of cilantro (fresh chopped)
1 tablespoon of olive oil
1 teaspoon of garlic (minced)
Salt

How to Make It:

In a large bowl, combine the 1/4 cup of lime juice with the coarse chopped chipotle chili in adobo sauce and dashes of salt. Toss in the 2 cut up boneless, skinless chicken breasts and cover. Refrigerate for 2 hours to marinate. Place the chicken pieces on greased skewers. Turn the grill to medium heat. Spray the 4 8-inch flour tortillas with cooking spray and grill them for about 45 seconds, flip and grill another 45 seconds. Place the skewered chicken and the pound of husked, rinsed tomatillos on the grill and turn every 30 seconds for about 5 minutes. Remove the chicken and tomatillos from the heat. Remove the skewers from the chicken. Chop the grilled tomatillos into bite sized chunks in a bowl. Add the tablespoon of olive oil and a dash of salt and toss.   Layer the tostadas by placing a tortilla down first, then divide the 2 cups of shredded romaine lettuce, tomatillos, chicken, 4 tablespoons of onions (fine chopped), and the 2 tablespoons of cilantro (fresh chopped). Garnish each plate with a lime wedge. Enjoy.

Italian Chicken

This savory Italian Chicken dish goes well with a salad or steamed vegetables. Makes 6 servings.

What You'll Need:

4 chicken breast halves (bone- in, skinless)
2 chicken thighs (skinless, bone-in)
1 can of tomatoes (diced, 15 oz.)
3 oz. of prosciutto (chopped)
1/2 cup of white grape juice
1/2 cup of chicken stock
1/2 cup of bell pepper (red, sliced)
1/2 cup of bell pepper (yellow, sliced)
1/4 cup of olive oil

1/4 cup of parsley (fresh flat leaf, chopped)
2 tablespoons of capers
1 tablespoon of thyme (fresh)
1 1/2 teaspoon of salt (divided)
1 teaspoon of oregano (fresh)
1 teaspoon of garlic (minced)
1/2 teaspoon of pepper

How to Make It:

Rinse and pat dry the chicken. Rub 1/2 teaspoons each of salt and pepper on all of the chicken. Pour the 1/4 cup of olive oil in a skillet and turn to medium heat. Add the chicken to the hot oil and brown on each side. Place on a platter and set to the side. In the same skillet add the 1/2 cups of chopped yellow and red bell peppers and the 3 oz. of chopped prosciutto and sauté. Stir in the teaspoon of minced garlic and cook for another 60 seconds. Add the 15 oz. can of diced tomatoes, 1/2 cup of white grape juice, 1 tablespoon of thyme (fresh), 1 1/2 teaspoon of salt (divided), 1 teaspoon of oregano (fresh), and 1 teaspoon of garlic (minced). Deglaze the skillet by scraping the bits from the bottom into the mixture. Add the cooked chicken and pout in the 1/2 cup of chicken stock. Turn the heat to high and bring to a boil. Cover, reduce the heat to low and simmer for about 25 minutes. When cooked, stir in the 1/4 cup of fresh chopped flat leaf parsley and the 2 tablespoons of capers, then serve.

Oriental Turkey Burgers

Here is a different twist to an American favorite, turkey burgers seasoned up with oriental spices. Makes 4 burgers.

What You'll Need:

12 oz. of ground turkey
4 hamburger buns (whole grain)
2 scallions (chopped)
1/2 cup of water (boiling)
1/2 cup of English cucumber (sliced thin)
1/4 cup of balsamic vinegar
1/4 cup of bulgur wheat
1/4 cup of yogurt (plain)
1/4 cup of cilantro (fresh whole)
1/8 cup of onion (sliced thin)
2 tablespoons of hoisin sauce
2 tablespoons of cilantro (fresh chopped)
2 teaspoons of canola oil
1 teaspoon of sugar (granulated)
1 teaspoon of ginger (grated)
1 teaspoon of chili garlic sauce
1/2 teaspoon of garlic (minced)
Salt and pepper

How to Make It:

First, combine the 1/2 cup of boiling water with the 1/4 cup of bulgur wheat in a small bowl. Seal with plastic wrap and set aside for about 50 minutes. Next, using a whisk in a separate bowl mix the 1/4 cup of balsamic vinegar with the teaspoon of granulated sugar. Toss in the 1/2 cup of thin sliced English cucumber and the 1/8 cup of thin sliced onions. Sprinkle with dashes of salt and pepper. Cover and set in refrigerator for half an hour. In another bowl, whisk together the 1/4 cup of plain yogurt with the teaspoon of chili garlic sauce and more dashes of salt and pepper. Set the bowl aside while preparing the turkey. When the bulgur wheat is ready, drain the water and add the 12 oz. of ground turkey, 2 chopped scallions, 2 tablespoons of hoisin sauce, 2 tablespoons of fresh chopped cilantro, teaspoon of grated ginger, and the 1/2 teaspoon of minced garlic. Mix with bare hands to insure good mixture. Separate into 4 patties. Add the 2 teaspoons of canola oil to a skillet and heat to medium high. Cook the turkey burgers until well done, 4 minutes on each side. Next, pour the cucumber mixture into a colander to drain, and then toss in the 1/4 cup of fresh whole cilantro leaves. Create the burgers by spreading the yogurt sauce onto each bun half, add the turkey burger, and then add a spoon of the cucumber cilantro mixture.

Enjoy.

Shepherd's Pie

This is a delicious and hearty one-dish meal. Makes 8 servings.

What You'll Need:

2 lbs. of potatoes (russet, scrubbed, peeled, chunked)
1.5 lbs. of ground beef (lean)
2 zucchinis (julienned)
8 oz. of mushrooms (button, sliced)
1 cup of carrots (peeled, grated)
1 cup of purple grape juice
1/2 cup of heavy cream
1/2 cup of bell pepper (red, julienned)
1/2 cup of Monterey Jack cheese (grated)
2 1/2 cup of beef stock (divided)
1/3 cup of butter
3/4 cup of onion (fine chopped, divided)
5 tablespoons of canola oil (divided)
3 tablespoons of flour (all-purpose)
2 tablespoons of tomato paste
2 teaspoons of Worcestershire sauce
1 teaspoon of paprika
1/2 teaspoon of cayenne pepper
1/2 teaspoon of garlic (minced)
Salt and pepper

Water

How to Make It:

Prep: Preheat the oven to 375 degrees Fahrenheit.

Add the 2 lbs. of scrubbed, peeled, chunked russet potatoes and the 1/2 teaspoon of minced garlic to a medium size saucepan and cover with water about an inch over the top of the potatoes. Dash some salt in the water and stir, turn heat to high and bring to a boil. Turn to medium high for about 15 minutes or until the potatoes are tender. Pour into a strainer to drain the water and return the potatoes to the saucepan. Add the 1/2 cup of heavy cream and the 3 tablespoons of butter and mash. Stir in dashes of salt and pepper and put aside.

In a large skillet add another tablespoon of butter with a tablespoon of canola oil and turn to medium high. Fry the 1.5 pounds of lean ground beef in the grease and add the 2 teaspoons of Worcestershire sauce and the 1/2 teaspoon of cayenne pepper and stir until the beef is well done. Season with dashes of salt and pepper. Add the 2 tablespoons of tomato sauce and stir over the heat. Pour in the 1/2 cup of beef stock and simmer for a few minutes. Pour the beef mixture in a large bowl and set aside. Add the remaining butter to the skillet and add 1/4 cup of fine chopped onions and sauté. Next add the 2 zucchinis (julienned), 1 cup of carrots (peeled, grated), 1/2 cup of bell pepper (red, julienned) and the teaspoon of paprika and stir. Cook an additional 10 minutes. Take off heat.

Create the pie by layering with 1/2 of the beef into the bottom of a 9x12 inch baking dish. Sprinkle the 1/2 cup of grated Monterey Jack cheese over the beef layer; add the remainder of the beef, pressing down firmly with a spoon back. Next, layer with the sautéed vegetables, and then add the mashed potatoes on top. Season the top with paprika and place in the oven to bake until the edges turn a golden brown, about 30 minutes.

While the pie is baking, make the gravy by pouring 3 tablespoons of canola oil into a saucepan and sautéing 1/4 cup of chopped onions and the 8 oz. of sliced button mushrooms, until they are soft. Stir in the 3 tablespoons of all-purpose flour and using a whisk pour in the 2 cups of beef stock and the cup of purple grape juice and stir until it thickens. Season with salt and pepper as desired. Pour over slices of shepherd's pie and enjoy.

Shrimp Scampi

If you love shrimp, you will love this recipe, complete with whole grain noodles. Makes 4 servings.

What You'll Need:

16 shrimp (large, deveined, shelled)
6 oz. of spaghetti noodles (whole grain)
6 black olives (pitted, chopped)
1/2 cup of onions (sliced thin)
1/4 cup of croutons (multi-grain, crumbed)
1/4 cup of parsley (fresh flat leaf, divided)
1/4 cup of chicken stock
1/4 cup of white grape juice
1 1/2 tablespoons of lemon zest (divided)
1 tablespoon of lemon juice
1 tablespoon of olive oil
1/2 teaspoon of garlic (minced)
1/4 teaspoon of red pepper flakes (crushed)
1/4 teaspoon of salt

How to Make It:

Cook the spaghetti noodles according to the directions on the package to "al dente." In a separate bowl, add the 1/4 cup of croutons (multi-grain, crumbed),

1/2 tablespoon of parsley (fresh flat leaf), and a tablespoon of the lemon zest and stir, let sit. Meanwhile, in a skillet, add the tablespoon of olive oil and turn to medium heat. Stir in the 1/2 cup of thin sliced onions, 1/2 teaspoon of garlic (minced), 1/4 teaspoon of red pepper flakes (crushed) and sauté for a minute. Stir in the 16 large deveined and shelled shrimp and the 1/4 teaspoon of salt and cook for another 90 seconds. Add the 1/4 cup of chicken stock, 1/4 cup of white grape juice, tablespoon of lemon juice and the 6 chopped, pitted black olives. Turn heat to high and bring to a boil, stirring and cooking for a minutes, then turn the heat back down to medium. Add the cook spaghetti noodles and the remainder of the parsley and lemon zest. Toss and pour into a serving dish. Sprinkle the 1/4 cup of crumbles multi-grain croutons over the top and serve.

Vegetable Pot Pie

Sometimes you simply do not need meat to make a full meal. This is a perfect tasty pot pie and all the better because it's homemade. Makes 8 servings.

What You'll Need:

2 pie crusts (9-inch deep dish, unbaked, rolled)
1 3/4 cup of vegetable stock
1 cup of carrots (thin sliced)
1 cup of English peas (frozen)
1 cup of potatoes (diced)
2/3 cup of milk
1/2 cup of celery (thin sliced)
1/2 cup of butter
1/3 cup of onion (fine chopped)
1/3 cup of flour (all-purpose, unbleached)
Salt and pepper
1/4 teaspoon of celery seed
1/4 teaspoon of garlic powder
Water

How to Make It:

Prep: Preheat the oven to 425 degrees Fahrenheit.

Place a saucepan over high heat and add 1 cup of carrots (thin sliced), 1 cup of English peas (frozen), 1 cup of potatoes (diced), and 1/2 cup of celery (thin sliced) and add enough water to cover the vegetables and bring the water to a boil. Add a lid and cook for 15 minutes, vegetables are done when they are tender. Drain water and set aside for a few minutes. Place a skillet on medium heat and add the 1/2 cup of butter and sauté the 1/3 cup of fine chopped onions. Add the 1/3 cup of flour (all-purpose, unbleached), dashes of salt and pepper, 1/4 teaspoon of celery seed, and 1/4 teaspoon of garlic powder and stir. Cook until well blended for a couple of minutes. Combine with the 1 3/4 cup of vegetable stock and the 2/3 cup of milk. Turn the heat to medium low and simmer for 5 more minutes. Turn off heat and stir in the cooked vegetables. Unroll a pie crust and place in a 9 inch deep dish pie pan. Add the vegetable mixture into the pie crust. Unroll the other pie crust and carefully place on top of the vegetable pie, sealing the edges but pressing a fork to make small ridges. Cut a couple of slits in the crust to vent the steam while cooking. Place on a baking sheet and in the oven for 35 minutes. Allow to sit to cool for about 10 minutes before serving.

Intermittent Fasting Diet Light Snack Recipes

Apple and Turkey Ham Salad

This is a delightfully crunchy sweet and savory salad. Makes 6 servings.

What You'll Need:

1/2 pound of turkey ham (thin sliced, torn)
4 endives (crosswise sliced)
3 apples (crisp, cored, sliced thin)
2 bunches of trimmed watercress
2 cups of onions (sliced thin)
1/4 cup of sour cream
1/4 cup of water
3 tablespoons of olive oil (extra virgin)
2 tablespoons of lemon juice
2 tablespoons of apple cider vinegar
2 tablespoons of Dijon mustard
Salt and pepper

How to Make It:

Add the 3 thin sliced apples into a bowl, pour over the 2 tablespoons of lemon juice, and toss to coat. Add the 3 tablespoons of extra virgin olive oil to a skillet on medium heat. Stir in the 2 cups of thin sliced onions and dashes of salt and sauté. Add the 2 tablespoons of apple cider vinegar and the 2 tablespoons of Dijon mustard in with the onions and stir with a whisk. Add the 1/4 cups of sour cream and water and continue stirring with the whisk. Pour the dressing over the lemon apples and toss. Toss in the 4 crosswise sliced endives, 2 bunches of trimmed watercress and the 1/2 pound of thin sliced and torn turkey ham. Add dashes of salt and pepper and toss before serving.

Baked Potatoes Twice

Baked potatoes are a tasty light meal, but "twice" baked potatoes are even better! Makes 4 servings.

What You'll Need:

4 potatoes (medium sized russet works best)
1/2 cup of onions (thin sliced)
1/2 cup of cream cheese with chives
1/2 cup of milk
1 tablespoon of butter

1 tablespoon of parsley (fresh chopped, plus 4 pinches)
2 teaspoons of thyme (fresh chopped)
1 teaspoon of canola oil
1 teaspoon of garlic (minced)
Salt and pepper

How to Make It:

Prep: Preheat the oven to 375 degrees Fahrenheit. Wash the potatoes, pat dry, and then rub the outside with the teaspoon of canola oil. Sprinkle salt over them and place in the oven, with a baking sheet on the rack below. Bake for 75 minutes; remove from the oven to cool.

Add the tablespoon of butter to a skillet on medium heat. Stir in the 1/2 cup of thin sliced onions and dashes of salt and pepper for about 7 minutes. Stir in the 2 teaspoons of fresh chopped thyme and the teaspoon of minced garlic, stirring for a minute. Remove from heat. Cut the potatoes in half, lengthwise, leaving the skin intact on the bottom. Carefully scoop the meat of the potato, leaving the skins intact. Place the scooped potatoes into the onion mixture and mix. Combine with the 1/2 cup of cream cheese with chives and the 1/2 cup of milk, the potatoes will be lumpy. Add the tablespoon of fresh chopped parsley and mix well. Evenly spoon the potato mixture back into the potato skins. Place the potato halves on the baking sheet and return to the hot oven for about 23 minutes. Garnish with a pinch of fresh chopped parsley and enjoy.

Broccoli Cheese Soup

Here is a lighter meal, made with wholesome broccoli, savory herbs, and delicious cheese. Makes 4 servings.

What You'll Need:

1 package of broccoli florets (frozen, 16 oz.)
3 cups of chicken stock
1 1/4 cups of Cheddar cheese (shredded, sharp)
1 cup of French bread (large cubes)
1 cup of onions (sliced)
1/2 cup of heavy cream
5 tablespoons of butter (divided)
3 tablespoons of flour (all-purpose)
2 tablespoons of olive oil (extra virgin)
1/2 teaspoon of garlic (minced)
1/2 teaspoon of thyme (fresh chopped)
1/4 teaspoon of white pepper (ground)
1/4 teaspoon of Creole seasoning
Salt

Nutmeg

How to Make It:

Add 3 tablespoons of butter to a medium saucepan turn to medium high heat. Sauté the 1 cup of sliced onions and add dashes of salt, nutmeg, and 1/4 teaspoon of ground white pepper. Stir in the 1/2 teaspoon of minced garlic and the 1/2 teaspoon of fresh chopped thyme for several seconds. While stirring with a whisk, sprinkle in the 3 tablespoons of all-purpose flour, keep stirring for about 2 minutes over the heat. Pour in the 3 cups of chicken stock, continue to stir with the whisk until all the lumps are gone. Turn the heat to high to bring to a boil while stirring. Turn the heat to low and simmer for 5 minutes, stirring often. Add the 16 oz. package of frozen broccoli florets and cook for another 10 minutes, stirring often. If desired, pour into a blender and food processor to blend. Or stir with a masher, mashing the broccoli. Return to the saucepan on low heat. Preheat the oven to 400 degrees Fahrenheit. Pour in the 1/2 cup of heavy cream, stirring while the cream heats. Pour in the 1 1/4 cups of Cheddar cheese, stirring until melted. Add the last 2 tablespoons of butter, stirring until melted and blended.

Put the 1 cup of large cubed French bread and toss with

the 2 tablespoons of extra virgin olive oil and the 1/4 teaspoon of Creole seasoning. Spread on a baking sheet and bake for 3 minutes in the hot oven. Remove to flip the croutons over and bake another 3 minutes.

Ladle soup and top with the croutons to serve.

Cauliflower Soup

Here is a delicious and filling, yet light, soup. Makes 4 servings.

What You'll Need:

1 head of cauliflower (chopped florets)
4 parsley leaves (fresh)
6 cups of chicken stock
1 cup of potatoes (scrubbed, skin-on, cubed)
1/2 cup of milk
1/2 cup of onions (chopped)
1 tablespoon of canola oil
1 teaspoon of butter
Salt and pepper

How to Make It:

Add the tablespoon of canola oil and the teaspoon of butter to a large saucepan over medium low heat. Stir in the 1/2 cup of chopped onions, cook and stir for 10 minutes. Add the heat of chopped cauliflower florets, 6 cups of chicken stock, and the cup of cubed potatoes and season with dashes of salt and pepper. Turn the heat to high and bring liquid to a boil. Turn heat to medium, cover and cook for 20 more minutes, until the vegetables are tender. Pour the soup into a blender or food processor to combine until smooth. Pour back into the saucepan and reheat to medium. Add more salt and pepper to taste. Thick soup may be thinned with extra milk. Garnish with a fresh parsley leaf in each bowl.

Greens with Baked Beans

This is a delicious one dish meal that offers wholesome beans along with smoked turkey ham and savory herbs. Makes 6 servings.

What You'll Need:

1 bunch of greens (mustard greens or Swiss chard, chopped, stems removed)
2 cans of pinto beans (drained, rinsed, 15 oz.)
1 can of navy beans (undrained, 15oz)
1 can of tomatoes (crushed 15 oz.)
1/2 cup of smoked turkey ham (diced)
1/2 cup of celery (fine chopped)
1/2 cup of carrots (fine chopped)
1/4 cup of parsley (fresh chopped)
1/4 cup of onion (chopped)
1/4 cup of water
1 tablespoon of olive oil (extra virgin)
1 teaspoon of garlic (minced)
1 teaspoon of thyme (fresh chopped)
1 teaspoon of oregano (fresh chopped)
Salt and pepper

How to Make It:

Prep: Preheat oven to 375 degrees Fahrenheit. Place a large skillet on the stove on medium heat and add the tablespoon of olive oil. Stir in and sauté the 1/2 cup of celery (fine chopped), 1/2 cup of carrots (fine chopped), 1/4 cup of onion (chopped), and 1 teaspoon of garlic (minced). Season with salt and pepper. Stir in the bunch of greens along with the 1/2 cup of diced smoked turkey ham and 1/4 cup of water. Cook for 3 minutes. Stir in the can of crushed tomatoes and turn the heat to medium high for 5 minutes. Stir in the 2 cans of drained, rinsed pinto beans, and the can of undrained navy beans. Next add the 1/4 cup of parsley (fresh chopped), 1 teaspoon of thyme (fresh chopped), and the 1 teaspoon of oregano (fresh chopped). Stir and heat through. Using a potato masher, mash the some of the beans (not all). Sprinkle with salt and pepper. Pour into a baking dish (2 quart) and cover with foil. Bake for 55 minutes, removing the foil for the last 10. Allow to cool for about 5 minutes before serving.

Maple Flavored Sweet Potato Fries

Here is a healthy and sweet version of a favorite, a nice alternative to French fries. Makes 6 servings.

What You'll Need:

5 sweet potatoes (peeled and cut into small wedges)
1 tablespoon of canola oil
1 tablespoon of maple syrup
1/2 teaspoon of lemon zest
Salt and pepper
Nutmeg

How to Make It:

Prep: Preheat the oven to 425 degrees Fahrenheit. Line a baking sheet with foil.

Place the sweet potato wedges in a large bowl, add the tablespoon of canola oil, and toss to coach each piece. Add dashes of salt and pepper. Place the wedges on the foil lined baking sheet and place in the hot oven for 20 minutes. Take the baking sheet out of the oven and place the wedges back in the bowl. This time add the tablespoon of maple syrup and toss to coat all. Place the potatoes back on the baking sheet and bake for 7 minutes, then flip the potato wedges and bake another 7 minutes. Add to a serving bowl, toss with the 1/2 teaspoon of lemon zest and dashes of salt, pepper, and nutmeg before serving.

Nutty Cucumber Mango Rice Salad

Enjoy something different made with peanuts, mangos, cucumbers and rice. Makes 6 servings.

What You'll Need:

2 scallions (sliced thin)
1 cucumber (English, diced)
1 jalapeno (red, seeded, diced)
1 1/2 cups of saffron rice
1 cup of mango (chopped)
1/2 cup of cilantro (fresh chopped)
1/3 cup of peanuts (salted, roasted, chopped)
1/4 cup of quinoa (rinsed)
2 tablespoons of lime juice
2 tablespoons of canola oil
1 tablespoon of lime zest
1 teaspoon of sugar (granulated)
Salt and pepper
Water

How to Make It:

Cook the 1 1/2 cups of saffron rice according to the package directions. In another saucepan add water and dashes of water and turn heat to high to bring to a boil.

Stir in the 1/4 cup of rinsed quinoa and turn heat to medium high. Cook for 12 minutes, until it turns tender. Pour into a colander and rinse with cool water and drain. In a separate bowl, combine the 2 tablespoons of lime juice, 2 tablespoons of canola oil, tablespoon of lime zest, teaspoon of granulated sugar and dashes of salt and pepper, using a whisk. Stir in the cook saffron rice, 2 thin sliced scallions, diced English cucumber, seeded and diced red jalapeno, cup of chopped mango, 1/2 cup of fresh chopped cilantro 1/3 cup of salted, roasted, chopped peanuts, and the cooked quinoa. Toss and season with more salt and pepper if desired.

Open Face Tomato and Mozzarella Herb Sandwich

This is a unique twist from a sandwich; there is no meat, just the delicious tomato and smoked Mozzarella with savory herbs on a baguette roll. Makes 4 servings.

What You'll Need:

4 slices of smoked mozzarella (thick)
4 slices of tomato (thick)
1 demi baguette (4 oz.)
2 tablespoons of parsley (fresh chopped)

1 1/2 tablespoons of Parmesan cheese (fine grated)
2 teaspoons of thyme (fresh chopped)
2 teaspoons of olive oil
1 teaspoon of garlic (minced)
Salt and pepper

How to Make It:

Prep: Preheat the oven to broil.

In a bowl, add the 2 tablespoons of fresh chopped parsley, 2 teaspoons of fresh chopped thyme, teaspoon of minced garlic and 2 teaspoons of olive oil and combine. Slice the baguette lengthwise in two. Cut in half so you have 4 pieces of bread. Evenly divide and spread the herbs over the bread, face up. Evenly sprinkle the 1 1/2 tablespoons of fine grated Parmesan cheese over the 4 slices. Bake under the broiler for 2 minutes. Remove and add a slice of tomato, and a slice of smoked mozzarella cheese on top of each slice of bread. Return to the broiler long enough for the cheese to melt, about a minute or two. Serve hot.

Orange Stir Fry Vegetables

This is a light dish made with just vegetables with the light fruity flavor of orange. Makes 4 servings.

What You'll Need:

1 can of water chestnuts (drained, 4oz)
1 cup of orange juice
1 cup of celery (chopped)
1 cup of mushrooms (rinsed, sliced)
1/2 cup of bell pepper (red, thin sliced)
1/2 cup of carrots (sliced)
1/2 cup of squash (sliced yellow)
1/2 cup of broccoli (chopped)
1/4 cup of baby corn
1/4 cup of snow peas
1/8 cup of onions (thin sliced)
2 tablespoons of cornstarch
2 tablespoons of orange zest
2 tablespoons of canola oil (divided)
1 tablespoon of soy sauce
1 teaspoon of ginger (chopped)
1 teaspoon of garlic (minced)
Salt
4 orange slices
Cooked rice (enough for 4 servings)

How to Make It:

Pour the cup of orange juice into a bowl and combine with the 2 tablespoons of cornstarch, tablespoon of soy sauce, teaspoon of chopped ginger, teaspoon of minced garlic and dashes of salt. Add a wok or large skillet to high heat and pour in the tablespoon of canola oil and sauté the cup of sliced mushrooms, 1/2 cup of sliced carrots, and the 1/8 cup of thin sliced onions for just one minute. Mix in the 1/2 cup of sliced yellow squash, 1/2 cup of chopped broccoli, and 1/4 cup of snow peas. Stir in the 4 oz. can of drained water chestnuts, cup of chopped celery, 1/2 cup of thin sliced red bell peppers, and 1/4 cup of baby corn. Cook for another couple of minutes. Pour in the orange sauce, stir and heat through. Serve over the cooked rice and garnish with the 2 tablespoons of orange zest and an orange slice on each plate.

Parsley Mint Roasted Carrots

This is a delicious way to eat your carrots, and nutritious to boot. Makes 4 servings.

What You'll Need:

2 1/2 cups of carrots (halved lengthwise and cut into 2-inch chunks)
1/2 cup of chicken stock
1/4 cup of mint (fresh chopped)
1/4 cup of parsley (fresh chopped)
4 teaspoons of olive oil
2 teaspoons of lemon juice
1/2 teaspoon of lemon zest
Salt and pepper

How to Make It:

Pour the 1/2 cup of chicken broth in a skillet and turn to medium high heat. Add the 2/12 cups of chunked carrots and a teaspoon of olive oil, stir, and bring to a boil. Place a lid on, reduce heat to medium, and cook for 13 more minutes. Remove the lid, stir and cook off all of the chicken stock and cook the carrots another 3 minutes to slight brown. Add dashes of salt and pepper. In a small bowl mix the 1/4 cup of fresh chopped mint,

1/4 cup of fresh chopped parsley, 2 teaspoons of lemon juice and 1/2 teaspoon of lemon zest. Place the carrots in a serving bowl and toss with the mint, parsley mixture. Serve warm.

Quinoa with Herbs

Quinoa is a super food because of the high levels of nutrients within it. Makes 4 servings.

What You'll Need:

2 3/4 cups of chicken stock
1 1/2 cups of quinoa
3/4 cup of basil (fresh chopped leaves)
1/4 cup of parsley (fresh chopped)
1/2 cup of lemon juice (divided)
1/4 cup of olive oil (extra virgin)
1 tablespoon of thyme (fresh chopped leaves)
2 teaspoons of lemon zest
Salt and pepper

How to Make It:

Pour the 2 3/4 cups of chicken stock into a medium size saucepan along with the 1 1/2 cups of quinoa and the 1/4 cup of lemon juice, turn to medium high heat, and bring to a boil. Place lid on saucepan, turn to low, and simmer for about 14 minutes. Meanwhile, combine the 3/4 cup of basil (fresh chopped leaves), 1/4 cup of parsley (fresh chopped), 1/4 cup of lemon juice (divided), 1/4 cup of olive oil (extra virgin), 1 tablespoon of thyme (fresh chopped leaves), 2 teaspoons of lemon zest, and dashes of salt and pepper in a small bowl. When the quinoa is cooked, add to a serving bowl and pour the "dressing" over, tossing to coat all. Add extra salt and pepper if desired.

Spicy Tomatoes and Green Beans

Sometimes you just need a pick-me-up and this dish will do it with the light flavor of cinnamon with tomatoes and green beans. Makes 6 servings.

What You'll Need:

4 cups of green beans (trimmed)
1 can of tomatoes (15 oz. crushed)
1 1/4 cups of water
1/4 cup of onions (chopped)
3 tablespoons of olive oil
Salt and pepper
Cinnamon

How to Make It:

Add the 3 tablespoons to a skillet and sauté the 1/4 cup of onions. In a medium saucepan, add the 4 cups of green beans (trimmed), 1 can of tomatoes (15 oz. crushed), 1 1/4 cups of water, sautéed onions, and dashes of salt and pepper and cinnamon. Stir and bring the water to a boil over high heat. Turn to medium low and simmer partial cover with a lid on for 35 minutes or until the green beans is tender. Season with more salt and pepper if desired.

Spinach Salad with Pomegranate Dressing

Spinach salad is a delicious meal made with pomegranate juice and walnuts, guaranteed to delight the taste buds. Makes 4 servings.

What You'll Need:

4 cups of spinach (baby)
1 cup of mushrooms (white button, sliced thin)
3/4 cup of tomatoes (grape, halved)
1/2 cup of walnuts (chopped)
1/4 cup of onions (thin sliced)
1/4 cup of pomegranate juice (plus 2 tablespoons)
1 tablespoon of apple cider vinegar
1 tablespoon of olive oil (extra virgin)
1 teaspoon of sugar (granulated)
Salt and pepper
Water and ice

How to Make It:

Pour 1/4 cup of pomegranate juice into a skillet and add the teaspoon of granulated sugar and a couple of dashes of salt. Turn heat to medium high and simmer for several minutes, stir often. Stir in the 1/2 cup of chopped walnuts and cook for another 5 minutes, the liquid should evaporate. Pour the nuts onto a cool baking sheet break apart when cooled.

Place the 1/4 cup of thin sliced onions in a bowl and cover with ice and water for 10 minutes. Drain the water and dry the onions with a paper towel. Put the 4 cups of baby spinach in a salad bowl, and then layer the cold onions, followed by the cup of thin sliced white button mushrooms, 3/4 cup of halved grape tomatoes, and the walnuts. In a separate bowl, combine the 2 tablespoons of pomegranate juice with the tablespoon of apple cider vinegar, tablespoon of extra virgin olive oil, and dashes of salt and pepper, using a whisk. Pour the dressing over the spinach salad and toss, and then serve.

Lightning Source UK Ltd.
Milton Keynes UK
UKHW021604130123
415295UK00016B/1324